Norbert Schwarz

Demographic

Rebuild

….after the War against Humanity, Beauty, Nature, Love and Life.

(author's translation without proofreading)

Druck und Distribution im Auftrag des Autors: Norbert Schwarz
tredition GmbH, Halenreie 40-44, 22359 Hamburg, Deutschland

ISBN
Softcover   978-3-384-36581-1
Hardcover   978-3-384-36582-8

Desouling of the World

This headline for the first chapter of a book on demography may come a bit unexpected. If you want to directly come to the point and read about demographics, you can certainly advance to the next chapter, where tables and graphs, such as the population diagrams or "population pyramids" will be displayed as you may expect when talking demographics.

The wounded Soul of a City

In my father's study, for a long time, hung a an art print of my hometown Frankenthal during the Thirty Years War, back then fortified as a star shaped fortress. Unfortunately, I cannot find this art print anymore. I was reminded of it, when I spotted the same art print in the corridor of a basement restaurant close to Frankenthal's cemetery. Today, nothing remains of the fortifications and the former star shape is not visible anymore, not even from above.

Today, Frankenthal is a rather insignificant small town of around 50,000 souls. Unfortunately, the town was not insignificant enough to remain spared from the air raids of the Second World War.

In a travel guide "Pfalz", I find the following introductory words about my hometown:

> *"The poet August von Platen praised Frankenthal in 1815 for being such a 'beautiful built town, one of the most beautiful places in the whole Pfalz'. Industrialisation in the 19th century and a devastating air raid in September 1943 have changed the face of the town. Today, Frankenthal is popular as a residential area for commuters to Ludwigshafen."(Schmitz-Veltin and Schmitz-Veltin 2011).*

About a year ago, a local department store put images and drawings of Old Frankenthal at the end of the 19th and the beginning of the 20th century on display; these gave you an idea of the 'beautiful built town' that Frankenthal once was.

Some old beauteous buildings or streets of houses still remain today, but two world wars in the 20th century, especially the Second World War have left their destructive marks. The last damaged buildings have been demolished in the 1970ies, so that I only know the completely rebuilt city. Apparently, a different city than the one described by August von Platen. Not a *"beautiful built town"* anymore and for sure not *"one of the most beautiful places in the whole Pfalz"*.

Why is the Modern World so Soulless and Ugly?

Why was it that as a student in Heidelberg I could learn so much better in the old town library (opened in 1905) than in the modern concrete and steel library in Heidelberg-Neuenheim (built in the 1970ies)? Why do I feel so much more at ease in old city centres than in modern ones? Why does the market square of Neustadt an der Weinstraße invite you to remain, but not the area around the main station of Ludwigshafen am Rhein? Is it about lack of money? Ludwigshafen had a steady influx of tax money from BASF, the largest chemical factory of the world.

Could it be that we are just not able to build as beauteous as our predecessors up to a 150 years ago were able to build? Could it be that we are a just not able to create a living environment ensouled and animated as our predecessors up to a 150 years ago were able to create? Allegedly we are looking back on 200 years of accelerated progress as never before in human history.

We also had two gigantic world wars in the 20th century. The Second World War brought an end to Old Frankenthal just as it had destroyed Old Ludwigshafen.

I leave it to academically trained historians to speculate about the causes of wars and their alleged geopolitical aims. The effects of wars are undeniable destruction of beauty and life.

The destruction of beautiful buildings, however, is not only happening during ongoing ("hot") wars. The main station in Altona was built in 1979 after the demolition of the predecessor building. An exhibition of pictures of the old station building in the station's underground crossings was a painful demonstration of the magnificent beauteous building that was lost with the demolition of the old Altona main station. It was replaced by a soulless and ugly concrete and steel structure. Today, we are not able to build such a magnificent station building as the old Altona main station. Apparently, already at the end of the 1970ies, we were not able to only maintain it, anymore.

So, destructions advance at a steady pace, from time to time accelerated by hot wars. It is as if we are dealing with a permanent war against humanity, beauty, nature, love and life.

Is Corona Better than War?

In spring 2020 I was still working as an infectious disease epidemiologist at a prestigious research institute and eagerly searched for data on cases of the newly emerging corona virus. This made me quickly realize that the course of the new viral disease was not exceptionally threatening, certainly not a killer disease (Markhoff 2020). Nothing that could justify the complete "lock down" that was about to get implemented. Nothing that required a vaccine intervention. Among colleagues I failed to convince with my data driven arguments; this was why I approached a supervisor with the appropriate epidemiological competence. Here, I learned that I should better care for our core areas and my projects.

Leaving no doubt about my professional assessment of the situation, I was seen off with the sentence "this is better than war". In that moment I realized that we were not dealing with a normal emerging infectious disease, but at least with political instrumentalization if not orchestration of the situation.

Not being able to find support among my scientifically educated colleagues, I went public by giving an interview; this drew to a close my occupation as a scientist and epidemiologist at this institute. For myself the "this is better than war" sentence allowed me to judge the coming events with a certain logic of war and act accordingly to live and to survive.

The corona-years did not leave any visible destructions such as bombed out houses. Nevertheless, there are consequences that could be effects of war:

- Excess mortality, especially as a consequence of side effect of the corona vaccinations. This excess mortality seems to persist throughout 2023 and 2024.
- Reduced fertility due to social isolation, but also due to spike toxic effects (Burkhardt, Lang, and Schwarz 2023), that reduce the probability of pregnancies and increase the probability of abortions
- Massive psychological traumata, that, just like the traumata of the world war generations are suppressed and collectively repressed

The decreasing number of births and the excess deaths have immediate effects on the population pyramid. However, these effects on the built up of the population are superposed by immigration. Immigration to Germany is on a particularly high level since 2015 without perceivable attenuation throughout the corona years.

War against Humanity, Beauty, Nature, Love and Life

What if the destructive decades of the last 120 years were a manifestation of a gigantic war against humanity, beauty, nature, love and life?

What if the powers that usually remain hidden in the background initiate wars with the aim of causing as much destruction as possible? Superficially, a war between groups of humans, e.g. different peoples, takes place due to conflicting geopolitical interests. These conflicting interests, however are nothing but a pretext for the war that serves for mutual and self-destruction of these peoples.

For pushing forward such destructive interests, it would not be of importance, which side wins a war, but rather that the respective war lasts as long as possible and is as destructive as possible. A long and destructive war requires military powers not to be too unequal, so it makes sense to support the weaker side. As we are (residually) ensouled human beings and able to feel compassion, we tend to sympathise with the weaker side. Our compassionate bent is being exploited by war propaganda without any scruples.

All in all, there are signs of a permanent war against humanity, beauty, nature, love and life. Such a war is in principle thinkable without acting entities fighting this war against us. Following psychoanalytical schools of thoughts, one could see collective manifestations of auto aggressive and self-destructive drives of us humans at work.

Due to my decades long academically-enlightened education (or indoctrination), it is hard for me to consider the possible work of divine or satanic entities as described in religious or occult circles. At the same time, I consciously decided, to also listen to and consider explanations, which are outside the dominating, materialistic view of the world.

Sober and empirical rational thinking makes me assert that there are godly and satanic powers, independent from the existence of a satanic or godly entity: If enough human minds are programmed to act along godly rules and form groups, which project political power, this de facto power can be regarded as a godly power. Likewise, we have to recognize the de facto existence of satanic powers due to the existence of satanic groups, independent of the existence or non-existence of satan.

In short, there is good and evil, nonregarding if non-human good or bad entities do exist or do not exist. God-evil polarity is used for propaganda purposes all the time, often in an abusive way to split up people and set them against each other to rage war and decimate one another. Creating concepts of an enemy through demonization is a standard war propaganda tool.

Demography is the science of populations referring to the old Greek word "demos". Other similar sounding words in old Greek are the word "demo", which refers to a demonstration and the word "demon", which refers to an evil spirit or an evil power.

I will use the word demography in the recognized sense with "demos" referring to the population that is being described with regards to its composition using statistical methods.

The German Volk – Shall We still exist in a few Generations?

Wars destroy our Volk (people)…. Can I write this? Shouldn't I use another word than Volk people or simply describe another Volk?

Robert Habeck, later minister of economics of the FRG, purported that the word "Volk" was Nazi-vocabulary. The public did not really appreciate being disavowed this way. To appease the people the publicly financed political-framing organization "Correctiv" pointed out, that the respective Habeck quote needs to be seen in the context of the question, how Habeck thought about the term "Volksverräter" (traitor of the people). The full quote was: "There is no "Volk" (people) and therefore there can be no treason of the people (Kill 2021).

This harmless appearing anecdote falls into a time, when the long-time resident population of our country – let's call them the autochthonous or indigenous population- after decades with low birth rates becomes aware of its own over aging. At the same time the biological decline of the population figures is being compensated by massive immigration, even over-compensated. Similar demographic developments happen in all ethnically European countries (Europe including Russia and America and Australia).

In principle, the phenomena I will describe in this book can also be observed in other European peoples. After all, the wars in Europe have devastated other peoples and I was tempted referring to population structure diagrams from other European peoples, for illustrating my observations.

The Italians or the Spanish for example are similarly affected by over aging as the Germans are. Our polish neighbour's fertility is, according to a quick internet search („Fertilität Polen") at 1.3 children per woman, thus far below the fertility of 2.1 required for natural population preservation.

In Germany, fertility was at 1.3 twenty years ago, and increased to 1.6 in 2021, mainly due to immigration of people of reproductive age with higher fertility figures.

If these developments continue the autochthonous or indigenous will inevitably become a minority of the population in European nations.

„Volk oder Bevölkerung" (the people or population)

Let us ask ourselves:
- Does the word „Volk" (people) has a value by itself in contrast to „Bevölkerung" (population)?
- Shall the Italian people still exist in a few generations?
- Shall the French people still exist in a few generations?

- Shall the Russian people still exist in a few generations?
- Shall the English people still exist in a few generations?
- Shall the Kurdish people still exist in a few generations?
- Shall the Irish people still exist in a few generations?
- Shall the
    - Yanomami
    - Tibetian
    - [any other people that comes to our mind]

people still exist in a few generations

And finally: Shall the German people still exist in a few generations?

You may have realized that I am struggling a bit with the translation of the German word "Volk". The closest translation according to dictionaries is people. Let us simply rely on our sentiment for words. We are the people! (Wir sind das Volk – this was a slogan of the 1989 protesters in Eastern Germany, shortly before the wall came down). According to a famous quote allocated to Kurt Tucholsky, the people do understand most issues wrong, but feel most issues right.

Let us continue using the word population for our now following demographic contemplations.

Population Preservation over Generations

Without im- and emigration, population growth is driven by births and population decline due to deaths. Demographers call this "natural population change". A good measure for fertility is the number of children a woman has birthed in average. For a natural preservation of the population this number has to be at 2.1. Since the 1980ies this fertility figure in Germany stagnated at 1.4 children per woman, with regionally lower numbers especially at the beginning of the 1990s in Eastern Germany, just after the collapse of the GDR.

Immigration of women with a higher fertility in comparison to the indigenous population, made the fertility figure increase to 1.6 before the onset of the corona-"vaccination" campaign.

With a fertility figure of 1.4 each generation reproduces itself only by 2/3 (1.4/2.1 =2/3=66.6%).

According to the population diagram of the year 1997 (https://service.destatis.de/bevoelkerungspyramide), there were 1,023,000 individuals (men and women) at the age of 25 years. A 25-year-old in 1997 belongs to the birth cohort of the year 1972.

Let us round this figure for our little thought experiment to 1 million 25-years olds in 1997 and let us assume a generation gap of 25 years. With these assumptions, 25 years later - in 2022 - we would have only 666,666 25-year-olds (1 million x 2/3) and another 25 years later - in 2047 – only 444,444 25-year-olds (Table 1).

De facto there were 953.000 25-year-olds in 2022. Thus, the low fertility was nearly entirely compensated by immigration.

Table 1: Hypothetical example of natural population declines of 1/3 from one generation to the next (25 years) with a fertility of 1.4 children per woman

BC = birth cohort	Number of 25-year-olds in our hypothetical example	Number of 25-year-olds according to Destatis*
1972 (BC 1947)		976,000
1997 (BC 1972)	1,000,000	1,023,000
2022 (BC 1997)	666,666	953,000
2047 (BC 2022)	444,444	926,000
2072 (BC 2047)	296,296	2070 901,000

* After 2022 Destatis projects the figures with moderate parameter assumptions (fertility 1.55; positive migration balance of 290,000 persons per year and live expectancy of 84.6 years for boys and 88.2 for girls); https://service.destatis.de/bevoelkerungspyramide

If the fertility remains low, autochthonous-indigenous inhabitants will become a minority among the 25-year-olds between 2022 and 2047. In 2070 less than a third of the 25-year-olds will not have a migration background.

These projections do not take into account the lower fertility due to the corona-vaccinations.

War and Demolition of Population Architecture

In war times more humans die than in peace times. Wars lead to over mortality. This over mortality is particularly high among men of military age, therefore a distinct surplus of women remains in the corresponding age-cohorts. Another demographic effect of war is a birth gap in years of war.

From the Population Pyramid to the Population Urn

When observing the current population diagram of Germany, the term "population pyramid" appears almost strange, as this does not look like a pyramid at all (https://service.destatis.de/bevoelkerungspyramide). Where does the term population pyramid come from?

Let us look at the population diagram of 1900 (**Figure 1**). This figure is from a publication on the population pyramid software GIZEH (Eicken, Lindemann, and Schulenburg 2004). While explaining GIZEH, the authors also explain some interesting changes in the composition of the German people over time.

The 1900 pyramid is easily recognizable as a pyramid **(Figure 1)**. Due to the fact that our life is finite, each birth cohort becomes smaller year by year. Increasing birth cohorts with increasing age cannot be brought about by natural population changes (birth and death).

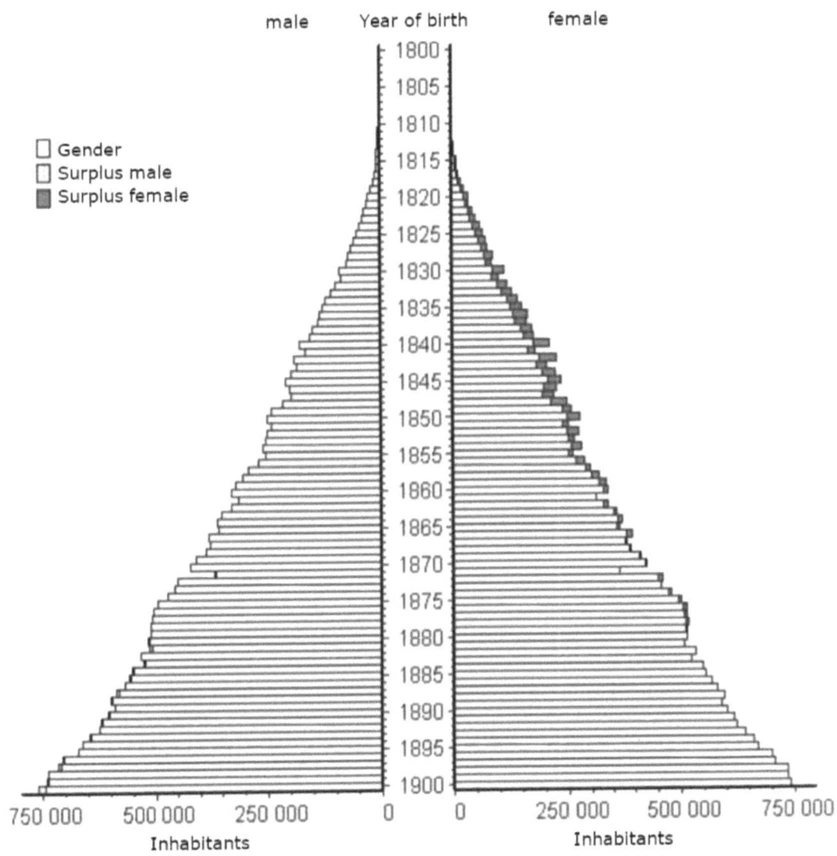

Figure 1: Classical population pyramid, Germany, 1900, census of the 2. December 1900 (Eicken, Lindemann, and Schulenburg 2004)

Birth Gaps due to Wars

In **Figure 1** years are listed in the middle as a labelling of the y-axis. Often, instead of years, age figures are given. The years are the years of birth of the respective birth cohorts. In a population pyramid of the year 1900, the 30-year-olds belong to the birth cohort of 1870. When looking at this year, we can see that the birth cohort of the immediately following year 1871 appears distinctly reduced for both sexes. Apparently, fewer children were born in 1871 than in the year before and in the following year.

So, what was going on in 1871? A simple internet research reveals that in 1871 the German French War lasted from July 1870 to May 1871. Many young men were away from home during that time, so that fewer children were born nine months later. This little retraction in 1871 is not dramatic yet and was quickly compensated by the birth cohorts of the following years. But let us assert: wars damage the population pyramid through over mortality and birth gaps. Over mortality in wartimes affects all age groups, although specifically men in their late teens and twenties are affected the strongest. War-caused birth gaps are concentrated on the birth cohorts of the war years themselves and are clearly visible in the population pyramid as retractions on both sides (boys and girls).

War Damages at the Population Pyramid

The Statistische Bundesamt in Wiesbaden offers a useful demographic tool allowing to show the population pyramids of 1950 to today and projected population pyramids up to 2070 using three different parameter sets (High, medium or low births, high, medium or low migration balances): https://service.destatis.de/bevoelkerungspyramide/.

Figure 2 shows the population pyramid of 1950. In this graph the central y-axis is labelled with birth years and age of the respective birth cohorts for both sexes.

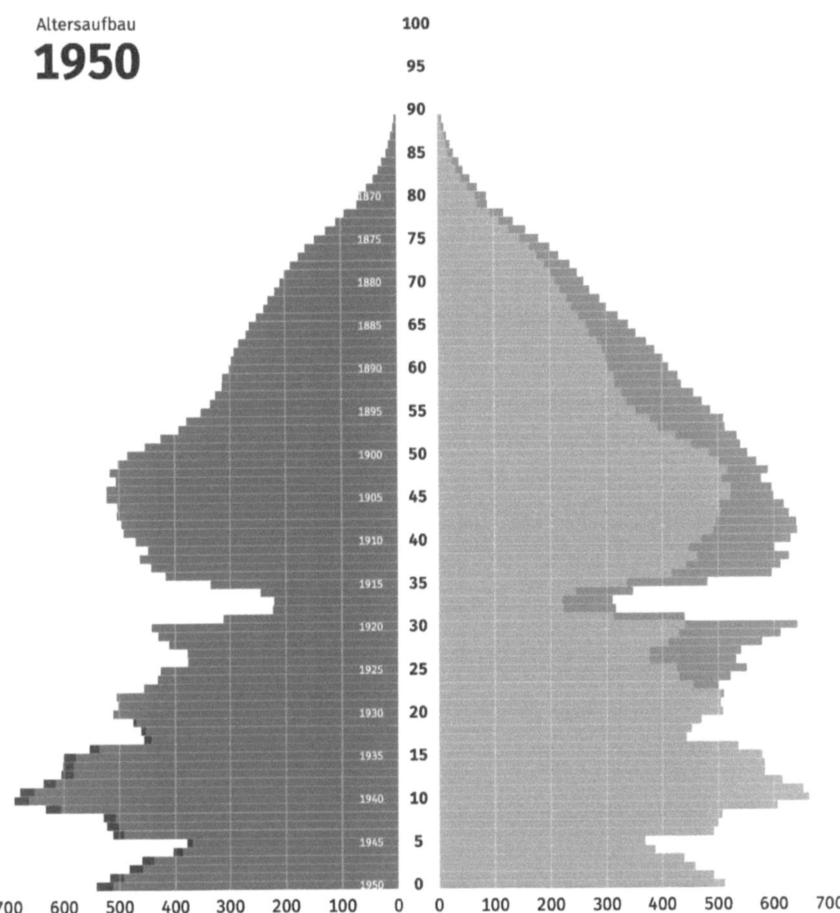

Figure 2: Population diagram of Germany in 1950 (https://service.desta-tis.de/bevoelkerungspyramide/#!y=1950&v=2)

For the years of the so called 2. World War the birth gaps of the years 1941-44 is clearly marked by retractions on both sides (never born boys and girls) followed by an even stronger retraction for the year 1945, which was the year of the total collapse.

Similarly, retractions due to birth gaps caused by the 1. World War can be seen in older age groups with the birth cohorts 1916-1918 most af-fected.

Table 2 shows the birth cohort figures for the years of the 1. and the 2. World War. Additionally, the proportional size of the birth cohorts in % of the pre-war year 1913 respectively 1939, are given. Birth cohorts of war years are distinctively smaller. Birth cohorts of 1917 and 1918 are only half the size of 1913. A similar picture can be found for the 2. World War. Here, the largest birth gaps with the smallest birth cohorts were found in 1945 and 1946. The immediate years after the 2. World War were probably characterized by plight and hardship among the defeated population. Noteworthy, these years appear to be excluded from history teaching in Germany.

Direct loss of human life in wars are best visible in the military age birth cohorts, where a surplus of women was brought about through mass killings of men in battle. (However, one also has to account for the 2-3 years longer life expectancy of women).

Table 2: Size of birth cohorts from 1913 to 1920 and 1939 to 1949 in Germany in 1950. Figures are from the population diagrams of the Statistische Bundesamt in Wiesbaden. (https://service.destatis.de/bevoelkerungspyramide)

Birth Cohort sizes in 1950 of the years around the 1. World War			Birth Cohort sizes in 1950 of the years around the 2. World War		
Birth-cohort	Number of individuals in 1950	% of 1913	Birth-cohort	Number of individuals in 1950	% of 1939
1913	1,055,000	100	1939	1,332,000	100
1914	1,013,000	96	1940	1,352,000	102
1915	817,000	77	1941	1,239,000	93
1916	592,000	56	1942	1,036,000	78
1917	531,000	50	1943	1,023,000	77
1918	539,000	51	1944	1,004,000	75
1919	752,000	71	1945	748,000	56
1920	1,084,000	103	1946	791,000	59
			1947	899,000	67
			1948	942,000	71
			1949	1,010,000	76

The example of the 1. and the 2. World War demonstrate that long lasting wars lead to birth gaps that become visible as retractions of these decimated birth cohorts in the population diagram.

The population diagram of the year 1980 **(Figure 3)** demonstrates a post-war normalization of birth figures in the years 1950 to 1965

”Baby Boomers" or rather a Temporary Normalization?

The birth cohorts between 1950 and the late 1960ies are often called the "baby boomers". But already a closer optical observation of the population diagram reveals that this "baby boom" is only a normalisation after the anomalies of the previous wars.

Not before the 1960ies, birth cohorts reached the magnitudes of birth cohorts before the 2. World War. The largest post-war birth cohort - that of the year 1964 (16-year-olds in 1980) - was already enlarged by immigration.

The 1964-birth cohort was 1,323,000 persons, slightly smaller than the birth cohorts before and at the beginning of the 2. World War. The baby boomers thus were not the expression of a particular strong fertility, but rather of a normalization of fertility without war.

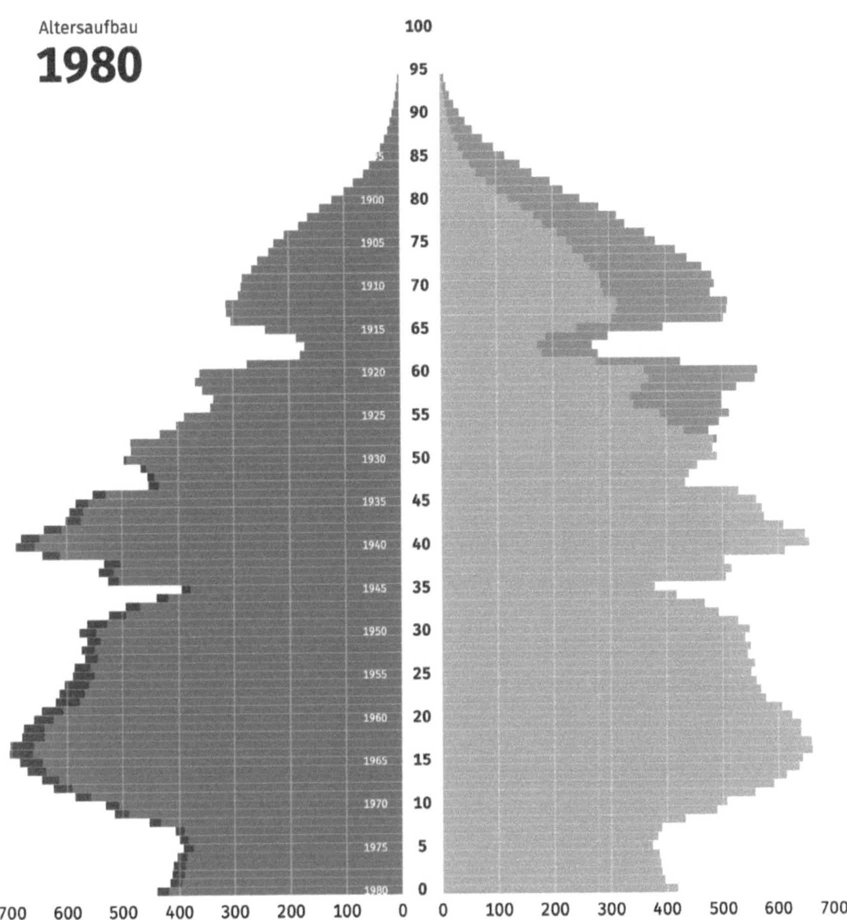

Figure 3: Population diagram of Germany in 1980 (https://service.desta-tis.de/bevoelkerungspyramide/#!y=1950&v=2)

The population diagram of the year 1980 (Figure 3) shows a sudden reduction of birth cohort sizes in the early 1970ies that reminds of the birth cohort reductions of the two world wars. Was there war again in Germany?

Falling Birth Rates from the 1970ies - War Again?

The declining birth rates from 1970 were labelled as a "Pillenknick – pill gap". In my opinion this trivializes the destructive societal phenomena taking place that are reflected by the birth gaps. I rather call this phenomenon by a term that may make you cringe: The Battle of Sexes. Let's face it: For around half a century men and women do not go along together well anymore. No wonder that there are no more babies.

The persistence of the birth gaps that set in from 1970 can be seen in **Table 3a**. The birth cohort of 1964 from the population diagram of the years 1997 is used as a reference year before the Battle of Sexes.

Table 3b allows an immediate comparison of birth cohorts of the two world wars from the population diagram of the year 1950 from Table 2 with birth cohorts from 1970 onwards that were decimated through the Battle of Sexes.

Table 3a: Birth cohort sizes in the Battle of Sexes in 1997 respectively 2017, from some representative birth cohorts (1977, 1987, 1997, 2007 and 2017) compared to the birth cohort of 1964

Birth cohort sizes in the Battle of Sexes in 1997 respectively 2017 https://service.destatis.de/bevoelkerungspyramide		
Birth cohort	Individuals in 1997 or 2017	% of 1964 birth cohort
1964 (33-year-olds)	1,482,000 (1997)	100
:	:	:
1977 (20-year-olds)	888,000 (1997)	60
:	:	:
1987 (10-year-olds)	945,000 (1997)	64
:	:	:
1997 (0-year-olds)	811,000 (1997)	55
:	:	:
2007 (10-year-olds)	736,000 (2017)	50
:	:	:
2017 (0- year-olds)	786,000 (2017)	53
Bonus information: birth cohorts 2019-2023		
2019	778,090	53
2020	773,144	52
2021	795,492	54
2022	738,919	50
2023	693,000	47

To allow for an easy comparison between the birth gaps due to the Battle of Sexes and the 1. and 2. World War, **Table 3b** contains the corresponding figures from Table 2 and Table 3a.

Table 3b: Birth cohort sizes in 1997 or 2017 from some representative birth cohorts in the battle of sexes (1977, 1987, 1997, 2007 and 2017) in comparison to the birth cohort of 1964. To allow for an easy comparison to birth deficits due to the 1. and 2. World War the size of the birth cohorts of 1913 to 1920 and 1939 to 1949 in 1950 are shown in grey.

Birth cohorts, Battle of Sexes, 1997 (2017 for birth cohorts 2007 and 2017)			Birth cohorts of the years around the so called 1. World War in 1950			Birth cohorts of the years around the so called 2. World War in 1950		
Birth cohort	Number of individuals 1997/2017	% of 1964	Birth cohort	Number of individuals in 1950	% of 1913	Birth cohort	Number of individuals in 1950	% of 1939
1964	1,482,000	100	1913	1,055,000	100	1939	1,332,000	100
:	:	:	1914	1,013,000	96	1940	1,352,000	102
1977	888,000	60	1915	817,000	77	1941	1,239,000	93
:	:	:	1916	592.000	56	1942	1,036,000	78
1987	945,000	64	1917	531,000	50	1943	1,023,000	77
:	:	:	1918	539,000	51	1944	1,004,000	75
1997	811,000	55	1919	752,000	71	1945	748,000	56
:	:	:	1920	1,084,000	103	1946	791,000	59
2007	736,000	50				1947	899,000	67
:	:	:				1948	942,000	71
2017	786,000	53				1949	1,010,000	76

Basic Principle of War: Divide and Set up Against One Another

History books report on wars as conflicts between peoples that were won by one side and lost by the other. De facto wars are always a heavy bleeding for the peoples that were set against one another. The real winners are usually third parties in the background, such as lenders, creditors and financiers who financially support both sides or deliver weapons. Human suffering and big losses of life, beauty and well-being and from a demographic point of view large birth gaps are simply accepted as part of the game. Every war is characterized by the division of humans, who are set against each other with the consequence of population reduction on both sides. This population reduction is to a large part due to birth gaps during the war years. Due to their geographic proximity wars tend to take place between neighboring peoples as currently the Russia-Ukraine war.

To accomplish birth gaps, it seems only consequential to divide and set against each other the humans who should be closest to each other and make children: men and women.

Persisting Birth Gaps in Assumed Peace Times

The two World Wars of the 20th century left vast destruction among others by devastating the soulful architecture of old cities. At least the birth gaps during the wars were made up for relatively quickly after cessation of violent battle action.

The normalization of birth figures after the 2. World War was called "baby boom" by demographers. Hot wars are staged along propagandistically created lines of divisions between peoples. The long-term sustained destructive effect hot wars have on the demography are through mass killings (of mainly young men in their twenties) and birth gaps. The never born of the decimated birth cohorts are later missing as parents.

But now, since the 1970ies, we observe birth gaps in assumed peace times as they could previously only be observed in war times.

The Battle of Sexes Decimated 10 Times more Birth Cohorts than the 2. World War

These birth gaps in assumed peace times are due to potential parents never becoming parents. As men and women simply do not find together anymore, they consequently get no children and do not start families. Such birth gaps can be asserted since the early 1970ies up to today and turned the former pyramidal shape of the population diagram into an urn-shaped ruin, which becomes clearly apparent around the year 2000 (**Figure 4**). Since 1972 the absolute number of births are roughly as low as during the war years 1942-1944.

Since around the year 2000 the number of births is even lower than it was in the year 1945, the year of total collapse.

While the number of births recovered relatively soon after the 2. World War, birth numbers remained low over five decades up to now with a tendency to even lower figures in recent years. The 2. World War has decimated around 4 birth cohorts. The Battle of Sexes since the 1970 has decimated around 50 birth cohorts up to now

The Population Diagram of 2002 – a Picture of Destruction

If we want to view the population diagram of 2002 (**Figure 4**) as a "pyramid", we need a lot of imagination only to perceive the ruins of a once magnificent building. The emerging form resembles more and more that of an urn, although demographers are still talking of a "population pyramid".

The population diagrams of the years around 2000, the turn of the millennials still contain the birth gaps of the 1. World War, while the generations that still contained the surplus of women of the 1. World War have already died off. The birth gaps and the surplus of women in older birth cohorts are still clearly visible.

In human populations all around the world, around 5% more boys than girls are being born. The resulting male surplus was enlarged by immigration from the 1960ies onwards, as usually more man than women migrate to our country.

From the late 1960ies, early 1970ies the massive birth gaps commonly known as "Pillenknick – pill gap" sets in that remains until today and which I see as a consequence of the Battle of Sexes.

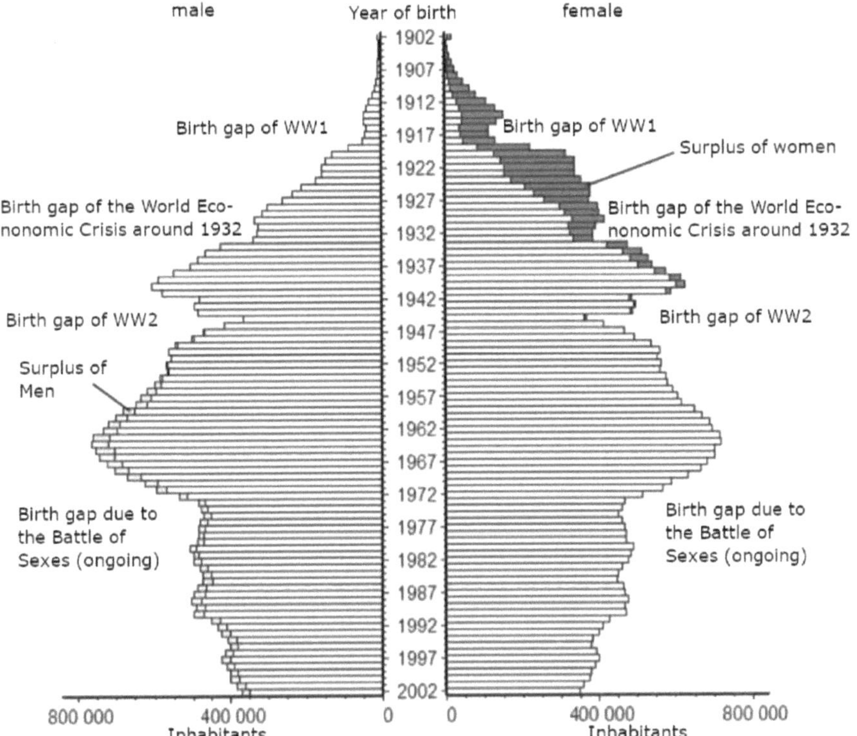

Figure 4: Population diagram of Germany in 2002 (Eicken, Lindemann, and Schulenburg 2004). The authors labelled the visible destructions using the terms commonly used by demographers.

With such Demographic Predisposition Taking any Part in a Hot War is Forbidden

Currently German media fosters a debate about the "Wehrfähigkeit", which means the preparedness for war with reference to the rather weak military power of the German Army.

Warmongering is mainly focused towards Russia, yet again. Non regarding how such a war would end from a militaristic-strategic point of view: The loss of human life in already small birth cohorts and an additional birth gap blow to coming birth cohorts due to the absence of the fighting young men would probably mean the end of the German people. Even a war fought on foreign territory could be deleterious to the already reduced autochthonous indigenous population, whose young men (and women) would be sent to getting killed.

We, the people cannot win a future hot war.

Battle of Sexes for Population Reduction

Let us look at a few aspects of the Battle of Sexes and how, since the 1970ies to today, it caused birth gaps of a magnitude only known from the World Wars of the 20th century.

Promotion of Non-Reproductive Forms of Sexuality

Matrimony between man and woman and the emerging family as the reproductive cell of people and society have been downgraded in comparison to other, non-reproductive forms of partnership. Non-reproductive forms of sexuality, such as homosexuality or the transgender perception are not only tolerated, but actively propagated. The explicitly expressed aim is to oust the heterosexual partnership between man and woman as the societal norm, and downgrade it to only one of many diverse forms of partnership. This downgrading consists of constant ideological nudging, usually using protection of minority as a leitmotif.

Raising the rainbow flag even on flag posts of official Institutions is being celebrated in established media as a paradigm of virtue and tolerance.

But why, actually? Isn't tolerating your fellow human beings implicit? Why does this implicitness have to be celebrated by rituals such as raising rainbow flags?

If intentional or not: The cult of colorful diversity boosts forms of sexuality that are non-reproductive and do not lead to children.

Sperm Decline

A meta-analysis that was published in 2017, conclusively showed that concentration and number of sperms in ejaculates of normal men in North-America, Australia and Europe more than halved between 1973 and 2011. The decline was also continuing in the last decade of the analysis with no signs of levelling off.

In the meta-analysis, these men were labelled as „Western men". Such decline of sperms was not observed in other men ("Others") from South America, Asia and Africa (Levine et al. 2017). **Figure 5** shows the decline of the average sperm count in Western Men between 1973 and 2011.

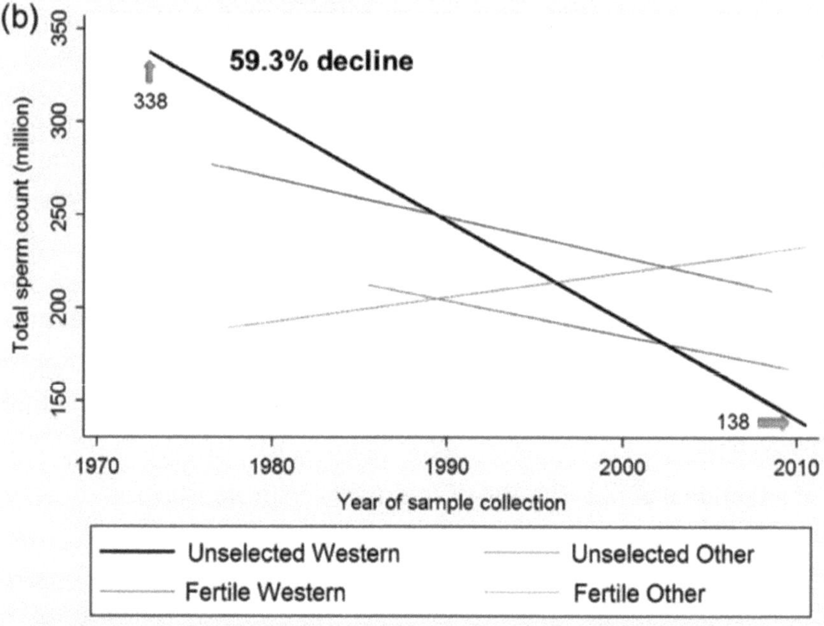

Figure 5: Results of a meta-analysis of studies on sperm counts in ejaculates over the decades between 1973 and 2011. In Western Men (European, North-American and Australian) a permanent decline of the number of sperms was found. In 1973 the average number of sperms in an ejaculate was 338 million and declined about 60% to 138 million in 2011) (Levine et al. 2017).

Fertility of Women and Age

Between 20 and 24 years of age women are most fertile and thus have the best chances (86%) for a successful pregnancy (Figure 6). Fertility then declines with age and usually ends with menopause in a woman's fifties. This age dependency of fertility is natural and well known, and therefore can be taken into account for life planning.

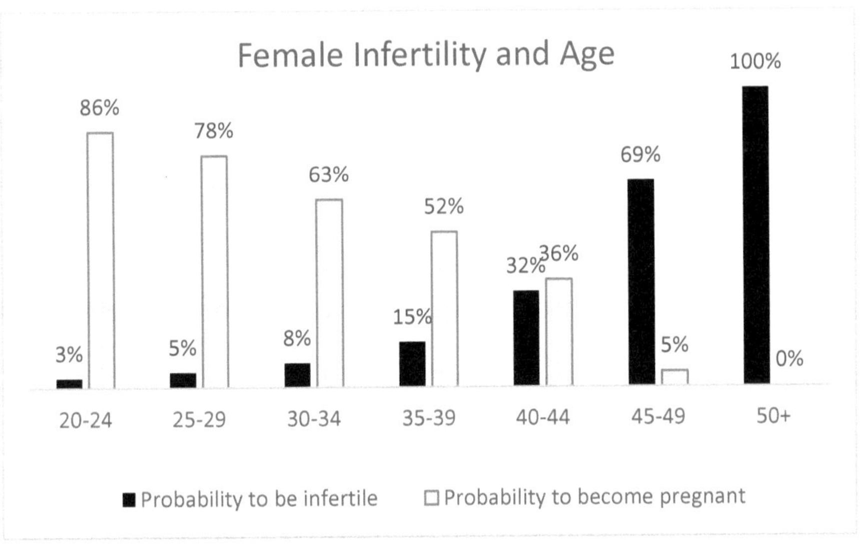

Figure 6: Correlation between age and declining fertility in women between 20 and 50 years of age. This graph was based on data from "Fertilia.de".

First Child with 30 Years of Age

The average age of the mother at birth of her first child in Germany in 2020 was 30 years. Fifty years ago, in 1970, the average age of the mother at birth of her first child was 24 in the FRG and 22 in the GDR (Statistisches Bundesamt 2020). Currently, Italian and Spanish women are in average even 1.5 years older than in Germany, when getting their first child. The fertility graph in Figure 5 shows that among 30–34-year-old women a wish for a child will only become fulfilled in 63%, implying that around 1/3 remain childless, even when trying hard. The probability of more children certainly declines with rising age of the mother at the first child

Work and Education keep us busy

We need money to live our lives and to make money we have to work and to find lucrative work we need degrees and certified qualifications.

Certainly, we should also enjoy our work and be interested in the content of our educational and professional course. Having a fulfilling work is nothing bad and as a society we benefit from motivated and diligent fellow human beings.

However, the prioritisation of education, work and career keep man and women away from finding each other, start a family and reproduce.

Long Educations Obstruct the most Fertile Years

The median age for graduating from universities in Germany is 25.4 years. This means that half of the graduates are younger and the other half older than 25.4 years. The median age when receiving a doctorate's degree is 30.5 years. After graduation starting into working life takes energy and time (job search, probation period, job changes) delaying onset of stability in life (settling down). Moving to a job may separate couples and endanger existing relationships.

Already the 30th birthday approaches, which is not celebrated with family and children anymore, but rather resembles the 20th birthday with friends and colleagues; or after several relocations, societal atomization and corona isolation takes place all alone without any friends.

When researching the age of university graduates in Germany, I noticed that the corresponding articles were all focused on career entry without even taking into account accompanying issues of life such as starting a family. As if there should be no more life apart from working life! Even an article with the promising title "Berufseinstieg mit 30: Bloß keine Torschlusspanik!" (Career entry with 30: Just do not fear being left on the shelf) did not even mention partner or children. Although the word "Torschlusspanik" in German always refers to the angst not to find a partner anymore, the article was only about explaining education times to a potential employer in a job interview (Campusjäger 2021).

Unfortunately, I could not find any figures on the age of non-academic career starters. However, with by now 56,4% of a birth cohort starting studies instead of taking on apprenticeship work, I can confidently make my point: Education times in Germany are too long and steal young women's (and man's) life time during their fertile years.

Work or University as a Place for Finding a Partner – Not Anymore!

In principle, places where we naturally spend long hours of our day together should be well suited to get to know each other in full authenticity. In a noisy nightclub, however, deep conversations are impossible and everybody plays an artificial, temporary role, making it a bad place for getting to know each other.

In the work place you naturally get relevant information about potential partners, for example when living as colleagues through work related situations that may challenge different character traits. Getting to know somebody at work also reveals the potential partner's ability to earn money and achieve resources for a future family.

Objectively, the work place is an excellent setting for man and woman to find each other as partners for life. Unfortunately, this has been made quite difficult over the last few years. Sometimes, partner finding opportunities are actively hindered by explicitly informing employees that romantic relations at work are not welcome. Additionally, fanatical protection against sexual harassment makes any relaxed interaction between men and women quite complicated.

Protection Against Sexual Harassment Makes Love Impossible

Recently, I had the opportunity to enter into the building of a county administration, which have become quite barricaded since the corona-years. I immediately noticed the central announcement board of which ¾ were covered with posters for diversity and against sexual harassment in the workplace. One poster pointed out that inappropriate words were already verbal sexual harassment and should be reported and prosecuted. I was granted only a moment, so I could not read all the posters in detail, but once thing was clear: They do not mess around, when it comes to protection against heterosexual sexuality.

Taking into account the over aging society that surely also affects local county administrations, I had some difficulties believing that sexual attraction played a major role at this work place. In rare instances, when a young man and a young woman should meet each other in this setting, the atmosphere of protective mistrust will squash the emergence of any romantic feelings.

If an inconsiderate remark can cost your work place, as a man you better keep a safe distance from you female co-workers.

Cooperation Turned into Competition

If a man and a woman get married, they wed each other. The German word for to wed –"sich trauen" – also means to trust each other. Additionally, "trauen" is related to "treu", which means faithful and loyal. It also translates to dare in the sense of being brave enough for taking a risk. The disastrously high divorce figures make you ask, if there is still enough trust between men and women to risk matrimony.

Humans competing with each other cannot fully trust each other. Competition impedes trust and loyalty. Real sustained love between competing human beings is impossible.

If the competitive battle of sexes is further fuelled, the traditional matrimony between men and women that once was the standard and the fundament of families may become a rarity. What effect on young men do job advertisements have that point out promotion of diversity and preference of women for employment and promotion?

Once the work place was an excellent setting for men and women to find each other. Having the same employer preselects for similar interests and life philosophies. In our atomized society often enough the work place is the only remaining place of human interaction. Unfortunately, the work place has become a minefield of political correctness and control, so that any deeper conversation has become dangerous. It may endanger your employment income, if you happen to have the (currently) wrong opinion. During corona games, expressing a slight understanding for critics of the government measures could get you fired. Meanwhile, questioning a politician's abilities and qualifications can be persecuted by law for "delegitimating the state".

Matrimony as an Existential Financial Risk

In case of a divorce both former partners face collateral costs that may turn the erstwhile safe harbour of marriage into an existential economic risk. Especially for men, getting married nowadays is from an economic point of view a high risk undertaking with hardly any economic benefits. "Don't get married" is the primary financial advice the American economist, podcaster and author Aaron Clarey gives especially to young men. His books carry titles such as „Bachelor Pad Economics" or "Life without the opposite sex". Such titles correctly imply, that Clarey tends speak out bitter truths, instead of soothing his audience with calming illusions. His secondary economic advice is "Don't get kids you cannot afford".

Destinies of Pay Dads as a Warning

I did not find a proper translation for the German word "Zahlvater" and decided to write "pay dads". The word shall imply a father who is obliged to pay monthly support payment for his children even if he has hardly any chance to stay in touch with them as contacts are often undermined by the mother reducing his fatherly role to regular cash transfers to her.

A lot has been written about the poverty risk, single mothers and their children face. The fact that single motherhood gets glorified in the media, as if it was a lifestyle choice will hardly improve the economic situation of single mothers and their children. For young men single mother glorification sends out the message: You are not needed and must be glad if a "modern strong and independent woman" tolerates you at her side (and the side of your children). Every man with an IQ above body temperature will be very cautious when it comes to relationships and in case of doubt abandon the venture of starting a family; economically there is nothing to win and a lot to lose for him.

The all-caring Nanny State on a closer look is not really caring about the needs of the single mother. Not if there is a father available, who can be forced to pay. The state primarily enforces support claims of the woman towards the father. Usually, the stipulation is measured with regards to the income. A family father who works full-time to gain resources for the family will be charged based on his full income, when it comes to alimony payments in case of a divorce. It seems nearly recommendable to cut down work hours if you fear a looming divorce. However, it can nevertheless happen that the stipulation will be oriented by the earning potential.

Instead of reducing working hours in the 50ies to have more time to enjoy life, the disposed pay dad then has to work full time to retirement age, in order to fulfill payments to his ex-wife and sometimes totally estranged children. This does not sound like an incentive to marry and father children.

Examples of divorce dramas bringing about financial ruin can be found in forum discussions, where men and women who suffered a divorce tell their own stories. I give you one example that seems representative enough to reflect the broader picture: *"All that's left from a middle-class family with a big house and a monthly net income of 5.500 € are three traumatized children and two hardship cases. As a former family father, I am still working, have 1180 € "Selbstbehalt" (the amount of money that remains untouched from third party claims) and my ex-wife has 960 € "Grundsicherung" (a guaranteed minimum basic social care). The house was sold off in a compulsory auction and is now being used as an accommodation for immigrants and homeless people. German family law produces only victims on both sides."*

Materialistic Weakening and Decimation

The quote "Mens sana in corpore sano"- a healthy mind in a healthy body is attributed to the Roman satirist Juvenal. Inversing this wisdom implies that an over-fatted, unhealthy body will also choke our cognitive potential.

In pre-corona times I enjoyed watching "Neues aus der Anstalt", a German political cabaret show and I remember the quote "Make the Germans rich and fat, then they will stop thinking". I am actually not sure anymore, who said this quote, could even be that it was not from this particular political cabaret format. However, it actually does not matter, who spoke this sentence. When just looking at the content of the quote, it very well describes the destructive principle of materialistic weakening. I would only remove the focus on the Germans, as materialistic weakening plays out on the whole of mankind, but especially on people living in economic prosperity and superfluousness; people, who are kept living in a sedentary lifestyle and who are fattened with carbo-hydrate rich food.

Indeed, apart from lack of exercise, it is actually carbo-hydrate rich foodstuff, that makes us fat and unhealthy. Carbo-hydrate rich does not only mean containing sugar, but also bakery and our "daily bread".

On Wikipedia one can easily find a list that ranks the countries of the world by their proportion of fat people defined as having a body mass index, BMI, over 30. This obesity ranking is headed by numerous small pacific Island Nations that had been blessed with Western prosperity food (Wikipedia 2024). The first larger nation on the list is Egypt on place 13 with 44.3 % of obese people, followed by Katar and Belize and finally the Western Imperial Nation USA on place 16 with 42.0% obesity. All over the world, 16% are counted as obese with a BMI over 30. Germany ranks on place 117 with around 20% obese people. The original source of this ranking is the WHO.

"Body Positivity" Reduces the Attraction between Man and Women

Well-trained masculine men are attractive to women. Likewise feminine, non-overweight women are attractive to men. Trivial, isn't it? Unfortunately speaking out such banalities can get you in trouble for "hate speech". Complaining as a man about obesity in women is "fat shaming" or a suppression of the "body positivity" movement. I did not have to translate these words as they are used in Germany without translation, thus pointing out, where these intolerant tolerance movements originated.

Praising obesity as "Body positivity" appeals to the laziness we all have inside. In contrast, developing healthy food habits, going outside every day, exercising and maybe s.t. going without something, for example without food (fasting) is hard work.

Special Heroes

The „body positivity" movement encourages me to be lazy and suggests that it is okay to follow lower impulses and guzzle junk food, whenever I want. Stay lazy as you are. It's a virtue! The German government issued a particularly bizarre example of inactivation propaganda, when they started the "special heroes" campaign during the corona games in autumn 2020. The special heroes campaign consisted of propaganda clips, showing young men and women, who did nothing but hanging around in front of the TV and stuffing themselves with junk food, thus being special heroes, who fight the corona virus by staying at home.

This way, a lifestyle is being promoted that shortens life-expectancy, reduces health and wellbeing and certainly lowers the chances for a man to find a woman and for a woman to find a man. Thus, the body positivity movement contributes to diminishing the probability for heterosexual couples to form. This falls in line with the promotion of non-reproductive forms of sexuality in that it reduces the number of children.

Women Become More and More Unhappy

Happiness research sounds like an area of science dedicated to contributing breakfast TV stories. But considering the importance of happiness for human well-being and that the „Pursuit of Happiness" is a central promise of the American Constitution, it may be worth looking at the results of happiness research.

Feminist ideology and the Battle of Sexes going along with it led to a stark decline of children being born with birth gaps of a magnitude only known from hot wars. Integration of women into working life is being celebrated as a big act of liberation. Women in western countries are in all political positions with material privileges. Often enough they are overrepresented by now and job postings in the public service never forget to point out that in case of equal qualification women will be preferentially hired or promoted.

And nevertheless women become increasingly unhappy since the 1970ies (Stevenson and Wolfers 2009). In the 1970ies American women were happier and more content than men. This difference became smaller in the 1980ies and in the 1990ies happiness parameters of women sank below those of men (Blanchflower and Bryson 2024). Certainly, it did not take long before somebody spoke about a "Happiness gap". The verbal proximity to another catch phrase, the "gender pay gap" may not have been a pure coincidence. For me as a man it sounds as me and my fellow males were to blame for this "Happiness gap", yet again. Another case of splitting the sexes. Interestingly, happiness and life contentment in the USA have gone down for both sexes over the last decades, only that the decline in women was much stronger than in men.

The corona games that started in 2020 then led to a worldwide collapse of happiness and contentment in both men and women.

Weakness Becomes a Virtue and Makes Dependent to the State

In the past, humans who suffered hardship and were in a situation of weakness were supported and saved by relatives or social contacts in the community. Such networks were the core family, the expanded family and the village community. In increasingly atomized societies such safety nets do not exist anymore and were replaced by institutionalized social welfare systems. Such, in principle, reasonable and necessary organizations have the side effect of creating anonymized dependency structure to the power apparatus that governs the allocation of funding and resources of the social welfare system.

Freedom and over all freedom of speech is something you need to be able to afford. If you want to dare claiming your freedom of speech, some savings in the back and a certain independence, when it comes to earning money helps. Critical statements about the corona measures mainly came from self-employed people or pensioners, who did not have to fear losing their job. The lesser you are able to gain your living, the more dependent on the social welfare system and the power apparatus you become. Weak humans are easier to control:

- Weak individuals are easier to submit and made dependant
- A population with only weak relational cohesion facilitates intrusive attacks on the individual
- Weak, lethargic and unmotivated broken people will not enter in long term heterosexual relationships and get fewer children

Evil Powers Thrive on Weakness

The term weakness may make you think of protection- and understanding -requiring forms of weakness. However, there are other forms of weakness, for example weakness of character or weakness to resist evil temptations. All these forms of weakness can be exploited by evil powers. If you are really powerful, you want to make sure that positions, from which political power is being exercised are filled with person you can control. Strong characters with honour and integrity will not be put into political positions of power. Having a past with corruption and criminal activities that were covered up are a predisposition for some position. Anything that makes the person vulnerable to black mail. Sexual activities that can be used for blackmailing may also be used for controlling politicians. Considering the importance of black mail for structuring ruthless power, narratives of elite gatherings with immoral and criminal activities to allow for mutual black mail seem somewhat plausible

The Four Generation Cycle

Hard times create strong men. Strong men create good times. Good times create weak men. Weak men create hard times (**Figure 7**). You may have come across this meme in recent times. The popularity of this meme maybe an indicator for more and more people realising that our "Western Democracies" are currently in a state of economic decline, along with cultural and in the worst case a civilization decline.

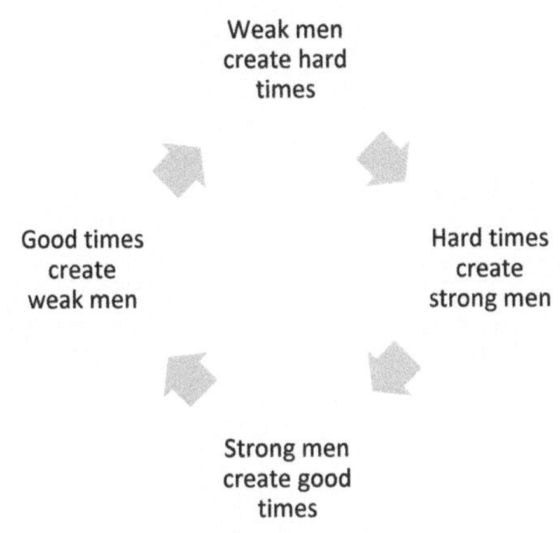

Figure 7: The 4-generation cycle of good times and hard times, strong men and weak men

Rivalry for Privileges instead of Competency Competition

Competition in a fair and meritocratic environment advances a society. However, with corrupt premises and low social coherence in an atomized society, a rather destructive rivalry for privileges and positions dominates. To what extent we find such phenomena in our country, for example in the public services, the universities or even in our ministries is beyond the scope of this little booklet.

Is our culture of competition really meritocratic in a way that would improve our life? Does meritocratic competition prevail making sure that important decisions are made by competent people? Did leaders in our country qualify for high and important position in a fair meritocratic competition? Or could they be the result of an unscrupulous opportunistic rivalry for privileges and favours? Rivalry for the favours of a ruthless power apparatus drains energy from useful competition. It corrupts splits society and fosters the divide and rule principle.

War Between Generations

A war between generations, between old and young is also tangible in all modern "Western Nations". Splitting old and young, like splitting men and women, also contributes to the decay of healthy social structures with families and communities. In Germany this war of generations was further aggravated by the Nazi era that was perceived as a morale-cultural downfall. The allies and their media further fostered the narrative of the Germans being people of perpetrators ("Tätervolk"). The guilt for horrible crimes that were without any doubt committed during the Nazi era was imposed on every single German human being. Following generations were encouraged in schools to constantly reproach their parents, and grandparents with regards to the Nazi era. Cautious and slightly anxious plea against the allegations, claiming that things actually were different ("Aber so war es doch gar nicht") led to embarrassed silence. The younger generation, who themselves were blessed to be born late enough, took their parents anxious plea or silence as indicators for culpable involvement in Nazi era crimes.

Were Germans of my Grandparent Generation bad Humans?

During the years of my medical studies and training I had many contacts with individuals of my grandparent's generation, who at that time were old and prone to disease and therefore numerous in German hospitals. (Somebody, who was 20 years old in 1940 was 80 years old in 2000). Most old people, I met in hospitals were normal people with strengths and weaknesses of character, usually friendly and considerate with good manners and decency, not so much different to old people in other countries.

Nevertheless, you must assume that many human beings did commit crimes during the Nazi era. There are good and evil deeds, does this imply that there are unequivocal good and evil human beings?

Allocating human deeds as good or evil is not following natural laws, but depends on categories developed by human thinking. Human beings are simply children of their times and develop behaviour depending on incentives (e.g. appreciation, money) or disincentives (e.g. exclusion from their societal environment). Different views are selected against, for example by shaming or even eliminating a person, who is articulating ideas that are not in line with the system of power. The actual murder of an idea or the human being articulating it is usually not carried out by a person in positions of political power, but through stooges, who claim to act according to orders (outsourcing any responsibility). But why are people so obedient?

Human Evolution fosters Abuse of Power and Submission

Public-Relation (or propaganda) may explain the frustrating fact, that humans tend to submit so easily to power and authority, especially in Western Democracies of the 20th century, where media control of people have been perfectionated. However, I do think that the tendency of humans to submit to power is also a legacy of human evolution.

No other animal has such a long phase of total dependency on conspecifics as a human. During these years of total dependency – childhood-obedient behavior gets incentivized time and again: An obedient child gets rewarded by the parents with love, care and attention. The power that we submit to during the years of our childhood is usually a good power. To this behavioristic explanation of human obedience, we have to add an evolutionary explanation: The probability as a human to reach reproductive age is higher if a child willingly submits to the positive and well-meaning power of the parents or other care takers. A child in the distant past, who willingly submitted to parental orders such as to stay with the group or to keep a safe distance from lions, must have had a higher chance to reach reproductive age.

Systems of power exploit this natural tendency to conformity and obedience, often by harvesting the obedience of young people for the state in schools. Unfortunately, character traits that make good parents are not the same ones that smooth the way into positions of political power. Nevertheless, political leaders are often collated to parents or care takers. The infantilisation of society going along with this is a most welcome side effect for control and maintenance of power.

Common terms used by media for political leaders are 'father of the people', 'big brother', 'mother of the nation' or with less pathos simply "Mutti". The word "Mutti" is derived from "Mutter", which translates to mother, however the effect of the "ti" ending gets lost in this translation. The online dictionary dict.leo offered the translation "mummy", which for me always was the corpse of an Egyptian pharao. As German media used "Mutti" for Angela Merkel I do not think that the latter translation is appropriate.

Uprising of the Youth

The years of youth and adolescence are characterized by a high level of idealism and the urge to make a positive impact to the future. These fundamentally good human features with their embedded urged to change things often leads to rebellious attitudes in teenagers. Systems of power, especially with totalitarian tendencies use this energy and channel it into system-stabilising trails.

The natural urge to question and oppose things during the storm and stress years between 15 and 25 is commonly channeled against parents and grandparents (and thus away from political systems of power).

The performer Jan Böhmermann uses agitation time and again to endorse the apparatus of power in TV formats that are camouflaged as satire and entertainment. In the German collective memory remained the staging of a children's choir singing a mockery song about their grandmother modifying it off track from using pet names to outright guilt blaming.

(The original granny friendly song „Meine Oma fährt im Hühnerstall Motorrad - My granny drives a motorcycle in the chicken house" was modified to close with „meine Oma ist eine alte Umweltsau -my granny is an old environmental hog". This story sounds harmless; however, it fits very well into efforts to split the generations using climate change blame narratives that are promoted with millions of Euros for example via the "Fridays for future movement).

„Astroturfing"

Organizations that endorse systems of power are always keen on winning young people for the "good common cause", which happens to be power conform (for example climate). However, the older generation should resist from claiming to having had a glorious past as a true rebel in their youth: Were the common causes of the past really genuine grass root movements? What if they were also media supported astroturfed control measures mainly taking place in the well-controlled environment of universities?

Nowadays young people are taking to the streets for demonstrations with good intentions for issues they (should) think are important. Independently if these issues are genuine or the consequence of propagandistic programming by media and schools, we should acknowledge the idealistic motivation, if we agree on the issue or not. We all are strongly shaped by societal programming. In my generation the atomisation of society and the decay of social and importantly family bonds became manifest. Apparently, we also failed to resist the social programming that is underlying these destructive developments.

For decades, the rebellious tendencies of young people are channeled towards older generations in particular towards the parents. At the same time system-endorsing resistance movements (sounds oxymoronic, because that's what it is) are founded and funded by super rich oligarchs or oligarchic clans.

Institutionalized social sciences programmed us to view the channelling of rebellious energies against parents as normal. The 5th of the 10 commandments to honour father and mother was being brushed off by psychologists and sociologist backed with the authority of science.

Certainly, one must also be able to afford critical remarks against an authoritarian regime. A retiree with savings from decades of economic prosperity and a regular and save pension does not risk losing the livelihood in form of a job in contrast to a mid-twenty youngster without any savings.

Freedom Movements that are Lacking Young People

In spring 2020, when the first resistance against corona measures formed, it soon became apparent that this resistance movement was mainly carried by over 40-year-olds. I personally experienced this phenomenon, when attending a demonstration in Heidelberg, a university town full of young students. As somebody in his late forties, I still belonged to the younger attendees of the event. Certainly, the proportion of young people in the general population is much smaller than it was 40-50 years ago, but even when taking this into account, young people were strikingly underrepresented on the demonstration.

Most Humans Want to be Good

The idealism of young people that was abused time and again throughout history at least demonstrates that most humans want to be good. Propaganda exploits this desire to be good by propagating action and behaviour that serve the powerful as being "good". If you then propagate behaviour that does not require effort, but meets the human laziness (such as staying at home), the obedience of the humans to be controlled is virtually programmable.

This strategy of suppression was regularly used during the corona games: The aim of the powerful was to split humans apart from each other and isolate them. Isolating yourself was declared to be ethically correct and virtual behaviour. To what frustratingly large extent this had worked has been already pointed out, often. However, I want to add the encouraging remark that this was only the perversion of a somewhat positive attitude of humans: The desire to be good.

Child Protection as a Pretext for Controlling the Parents

There are damaged families. These are to a certain extent the result of the war against humankind, beauty, nature, love and life. So, in some cases help from outside the family may be justified. Unfortunately, the child protection argument is also a perfect pretext for intruding into families, disempowering the parents and submit and control them. Good and competent parents have experienced this overreach of authorities during corona times, when wanting to safe their children from repressive mask mandates or the syringe madness.

One of my colleagues in the Reutlingen laboratory for pathohistological examinations of corona vaccine-victims experienced nannyistic control by local authorities after she and her husband had taken their children out of school during the corona games and taught them at home. Funny side aspect of this experience: All three children's performance in tests and school exams improved and soon overperformed their class mates, who had remained under care of the school system.

When a Powerful State Wants to Protect You

If you give up freedom for security, you will lose both. This quote was very popular on demonstrations against the repressive measures of the authorities and the powers in the background. When entering this quote into a search engine (May 2024), I find a lot of articles that try to frame the quote, mainly by telling the reader, that the quote was used by the wrong people (opponents of the corona regime, "Querdenker" etc.). Over the last 4 years, apart from protecting the population from "medical false information", the protection from psychological harm through "hate speech" was something global power machines were particularly concerned with.

Nothing should be written or said that could be offensive. Usually, there were no well-defined criteria, what could be offensive and thus "hate speech". Consequently, everything could be declared to be an act of "hate speech".

„Free Speech" Becomes „Hate Speech"

If you are personally offended, for example by a derogatory remark about your outward appearance, similar neuro-psychological reactions are being induced as in case of disagreement e.g. on political issues. If the power apparatus wants to protect you from such uncomfortable neuro-psychological reactions, disagreements, just as offences, need to be suppressed.

Or to phrase this the other way around: if we want to preserve free speech, it must be possible to expose somebody to uncomfortable neuro-psychological reactions. It must be possible to offend somebody. In a society with healthy social relations (e.g. families) natural mechanisms to cope with offences exist outside of state given laws. Good conventions and manners come to mind.

In Germany „Deligitimation of the state" was established as a criminal offence. This stretchable paragraph is now a constant threat to critics of politicians who now enjoy special protection against the uncomfortable neuropsychological reactions that offences and dissenting opinions could cause. This protects politicians from critique and strengthens their power.

Is Motherly Love Racist?

„Nächstenliebe" translates to altruism or charity according to online dictionaries. The literal meaning of the "Nächste" gets lost in this translation. "Liebe" means love. The "Nächste" can be translated to the next, which implies on one hand that you indeed ought to love the next one crossing your ways. But it also implies a preferential love for people, who are closest to you, your family members or some close friends. This preferential love for humans close to you gets contradicted by the deeply implanted angst to discriminate somebody.

Giving preferential treatment to your own child? Isn't that already racist? Its motherly love racist? Better confront your otherwise contorted and spoiled brat with the thitherto rarely brought up concept of diffidence and abstinence in demonstrative favouring treatment of the foreign child! Or do we have to offer our love without any discrimination to any stranger we come across? Do we instead of loving our next, first have to love the distant stranger?

Hospitality – a Worldwide Recognized Concept for Offering Love to Non-Relatives

Giving preference to your own family is an entirely normal behaviour pattern, worldwide. For non-family members you have hospitality. In Germany we have replaced hospitality by institutionalised welfare, that is characterised by money handouts instead of warm-hearted empathy. Needs for human relations are met in more and more parallel societies that do not live together but rather next to each other.

The cohesion of ethnic groups in a foreign country can be stronger than it was in their home country. This intra group cohesion can also lead to exclusion and in the worth case ostracism towards peoples outside this group for example those of the indigenous-autochthonous population. If such ethnic groups with a strong cohesion become powerful in the schoolyard, the rather individualistic raised autochthonous child may get in trouble.

Natural Grown Groups Work Better than World View Based Groups

In the past groups formed due to proximity in space. People in a village meet regularly and interact in a natural way. People who enjoyed talking to each other formed friendships. Sometimes there were disagreements and quarrels over clear lines of conflict, e.g. over border of overlapping properties and certainly people did not have the same opinion about everything, but a good naturally grown friendship could survive such disagreements.

When a man and a women found each other, married and got children, a new small natural group emerged that was embedded into a larger network of relatives, the extended family.

The repressions of the corona regime in 2020 created additional disruptions in an already damaged society with weak families and weak overall social cohesions. Corona propaganda created conflicts that caused families to fall apart, destroyed marriages and divided friendships.

Purely World View Based Groups are Unstable

At the same time world view-based groups emerged quickly among the opponents of the oppressive corona regime. Societies and political parties were initiated and countless lose interest groups formed in electronic media. Certainly, it was a relief to realize not being entirely alone. Especially the virtual groups on electronic media may have facilitated to establish countless contacts (or to give you the illusion of having these contacts), however such electronic contacts rarely bring about long-term stable relations.

Besides, world view-based groups with a clear oppositional direction were an easy target for corruption through moles acting on behalf of the power apparatus. We have to assume that such moles were involved in every arising oppositional group and some groups may have been directly founded by the power apparatus.

During our scientific work on corona vaccine damages, I had to experience that the most impetuous attacks on our work came from inside organizations that we shared interests with and who supported us.

Key Role of the Media

Who controls the media and the education system controls the people.

The war against humanity, beauty, nature, love and life reached its population reducing effects mainly through propaganda. Or to phrase it differently: School and media reprogrammed us in a way that we do not form couples anymore and do not reproduce anymore.

However, alternative media that mainly gained power due to the internet seem to become a game changer. Traditional media were concentrated in the hands of very few oligarchs and allowed nearly total vertical control. With the internet media, people can make their media out of the peoples. This makes a correction of the false programming, to which our generation was still fully exposed to, possible again. The desire of freedom that awoke due to corona repression thus may finally be key for demographic survival (some talk of a "Great Awakening").

Do We Live in the Best of all Times?

The quote "We live in the best Germany of all times" was used by the FRG president Frank Walter Steinmeier. It was written down in a commemorative speech script titled "We lucky children in the middle of Europe" given on the 3rd October 2020, a point in time, when corona measures had already made clear, how repressive this "best Germany of all times" actually was (Steinmeier 2020).

With this speech Steinmeier used a propagandistic technique that works out by strengthening the status quo preference bias that is embedded in us.

Status-Quo Glorification and Badmouthing the Past

In „Angst and Power" on propaganda techniques, the psychologist Rainer Mausfeld explains the status quo Bias: (Mausfeld 2019): It relies on the natural tendency, to accept the current condition of society as good, just, morale, legitimate and desirable. We have and embedded tendency to prefer the current status quo to all alternatives, even if they were objectively better. We are always prone to downplay disadvantages of the current status quo. This preference of the status quo as such is not a bad thing. Without this disposition we would probably be in permanent stress and could never reach the relaxing state of content. However, this status quo preference makes us involuntarily support suppressive powers and attack those pointing out drawbacks, as they appear to disturb our relaxed mental "everything is fine" state. The status suo bias is constantly being used in every political system, especially when things are actually not fine, but maintenance of power requires to make people think they are fine.

Depreciating the Past in Written History

Using the status quo bias for glorifying the present age is a crucial pillar of power. We have to assume that the powerful always had a high interest to glorifying the condition of their age. The easiest way to glorifying the presence is depreciating the past. The past is depicted darker and darker from generation to generation, thus creating a dark image of the past. This has become literal in written history with expressions such as the "Dark Ages".

Over the last 4 years as an infectious disease epidemiologist, I had a front row seat view on the corona games. Thus, I could observe, how current events in a field I am very familiar with were waved into (nearly truth-free) stories by mainstream media making it obvious that you simply cannot trust the official narration. But if the narration of the present is already falsified beyond recognition, how does it look like if we take narrations from the past?

History is written by the victors. In Germany, we surely had to experience this painfully, but not only we in Germany. Most people in the world do not have power and are subordinated to power, among others to the victors of history, who write the history.

A Prison of Mind – How are so Few Able to Rule over so Many?

Eighty individuals are in principle stronger than eight. Eighty million people are in principle stronger than eighty. Eight billion people are in principle stronger than eight hundred. The de facto power actually lies in the hand of the Many, who are nevertheless de facto subjugated to the power of a Few. The fact that very few rule over very many is known as "Hume's paradox" named after the Scottish philosopher of enlightenment David Hume (1711-1776).

What is keeping up the Hume's paradox is not factual power, but fictitious power. The power of narrations, the power of stories, the power of history. A prison of mind, with media, politicians and ourselves being prison guards.

Especially in states and power systems that need to keep up the illusion of granting freedom of speech, the media and opinion industries are of extraordinary importance for maintaining power and control: If you grant the subjugated to speak out freely, what they think, the controllers have to make sure that these thoughts are in line with the power maintenance requirements of the powerful (Herman und Chomsky 1988).

The function of the media machine lies in directing public opinion in a way that produces a power-confirm majority opinion. Furthermore, the mainstream media tries to instigate people with the desired opinion to discipline other people, with "wrong think" tendencies. This can be reached by defaming and ostracizing other opinions. How well this works was proven during the corona games.

How Much of our Written History can We still Believe?

The depreciation of the past as a power principle, implicates that the past must have been much better, than we are being told and taught. The ironic "Früher war alles besser - everything was better in the past" may contain a lot of truth. Probably, we were much better in the past (at least than we think, nowadays).

These insights also bring about hope: Apparently there is more good in us humans than we would concede to ourselves. Also, the insight that our lack of freedom mainly stems from fictional and not real power is encouraging, as it means that it is on us to free ourselves. For the collective, or all of mankind this certainly is easier said than done, but as an individual you can already increase your own degrees of freedom tremendously: Recognize the manipulative role of media and walk away from it.

Virtual Communities Cannot Replace Real Communities

Real social life takes place in real space. In the first chapter of this booklet, I remarked that the contemporary way of shaping our living environment, especially our architecture and town construction lacks soul compared to the past. Our demographic rebuild does not only encompass families and children for the past generation, but also the revitalization and ensouling of our habitats. Looking at old beautiful buildings makes you realise, what potential of creating beauty us humans do have. Such beauty in architecture of the past with splendiferous buildings can be found in buildings all over the world.

Ensouled Architecture Can be Found Everywhere in the World

Oddly enough, beautiful architecture and ensouled buildings can be found all over the world. However, everywhere you find the same pattern of cities with ensouled old towns surrounded by soulless suburbia for storing humans. The first time I realized this phenomenon being a worldwide one was in cities outside of Europe such as Isfahan and Shiraz in Iran and Samarkand in Uzbekistan.

However, the official histories of construction of all these buildings, especially the large and splendiferous ones seem questionable. Noteworthy, such buildings are usually occupied by institutions of the power apparatus, such as courts or political institutions. To consolidate power, it makes perfect sense to also adapt the history of such a building and all other eye-catching monuments to the requirements of power legitimation. Look at any German city that has not been reduced to ash and rubles entirely by air raids, and reflect on how the most beautiful buildings have come in to being.

The stories about English colonial powers having erected all buildings in royal colonies that are nowadays called "Victorian" appears at least questionable. The experience I made with English craftsmen does not really support the official history of Victorian architecture.

Demographic Rebuild

How is it possible to reconstruct our population pyramid and giving it a solid base, again? Or less academically overblown: How to make more children?

How to Accomplish the Demographic Turnaround?

How do you like the slogan "Let's dare more demography"? Sounds not too bad, however, this is still too much in the political sphere. ("Let's dare more democrazy" was the main slogan of the Willy Brandt government of 1969). Politics is always 'divide and rule' and it's the divisions we need to overcome now. Thinking in political parties and 4-years episodes is not made for knocking down the walls build in our heads.

We are talking about no less than a transformation of our public spirit. In a pure materialistic realm of thinking this means a reprogramming. The key for such a transformation lies in the education and media system and soon, hopefully, in the families again.

The Battle of Sexes decimated 10-times more birth cohorts than the 2. world wars. Every war is preceded by splitting the humans to be set against each other. For the Battle of Sexes the human beings that were supposed to come together, build families and get children were set against each other: men and women.

How can this divide be overcome?

Love is the Answer

You may think that the statement in the header is quite trivial, when reflecting on raising the birth rate. You may have expected something a bit more complex; something more "sciency"

I do not object. Actually, there is no more reason for me to act out sciency rituals, as I was kicked out of the scientific establishment 4 years ago for criticizing the corona measures.

Once your reputation's shot you get away with quite a lot!

The destructive "Follow the Science" campaign of the corona games and its effect on millions of people's health should be a warning for the future, not to blindly put authorities on a pedestal, only because somebody put the label "scientific" on them.

But, why can't we put good old-fashioned love between a man and a women with the resulting family back on the pedestal?

This is no ground-breaking, elaborate innovative ideal. The love between man and woman and their family is deeply engrained in us; and it is a central leitmotif in legends, stories and lived traditions. It actually is the leitmotiv of human life as such and its perpetuation.

For my generation (1970ies birth cohort) the demographic turnaround is certainly not attainable anymore. Apparently, it was possible for schools and media to brain wash our generations in a way that we simply do not form sustainable couples and long-term partnerships anymore. Consequently, our generations hardly reproduced. If such a brain washing that is against human nature can be attained in only a few decades, it must also be possible to raise future generation in line with our nature to normal men and women, again.

Mindset of a Forester

Rotation time in forestry is the average time between planting and harvesting trees. For spruces this rotation time is around 100 years, for pine trees around 120 years, for fir trees and beeches around 140 years and for oak trees around 200 years. Only a small fraction of the forest was planted during a forester's lifetime and a tree that is mature for harvesting had been planted many generations ago. At the same time a forester does not see any of the self-planted trees grow to become a mighty tree during his life time. If the biodiversity of a forest declined, because for several generations only spruce trees were planted, it's nevertheless worth turning around the decline so that healthy mixed forests will thrive, when the forester is long gone. A certain connectedness between generations seems to be inherent to the forester profession.

Demographic rebuild with reawakening of the love between men and women and families with thriving vivacious children revitalizing our beautiful villages and cities is a task for generations, just like the profession of a forester.

Us the generations of the decimated birth cohorts will certainly only experience the beginning of such positive changes in our lifetime.

The German East-West Unification Worked Well on the Human Side instead of 25 Years of 'Divide and Rule' by Media and Politicians

But is it really possible to overcome the deeply ingrained indoctrination that has left such a wide split between men and women?

If we look at other splits between human groups, we can hope that this will be possible. During and after the reunification of East and West Germany, media never ceased to stress differences between the "Wessis" and "Ossis" as if trying to keep up the split. If you keep in mind that the 1960ies and 1970ies birth cohorts were indoctrinated to shoot your fellow Germans in case of a hot war between East and West it should not be so difficult to keep up the split. No, doubt that the established media keep on doing so. However, if you travel our country or look at for example workplace environments, East and West Germans have grown together. Wir sind ein Volk! (We are one people – also a slogan of the 1989 protesters in Eastern Germany, which was derived from "Wir sind das Volk" – we are the people).

For the demographic rebuild the birth cohorts from the 1960ies and 1970ies will play no role. Young adults now are also programmed in a way that makes an immediate demographic turnaround seem unlikely. However, non-regarding if you appreciate the engagement of younger generation for a common (usually astroturfed) purpose, such as "climate", this engagement shows the deeply embedded will of young people to be good. Now, it is on us -the people- to take the media and education system and restructure it. The embedded urge to be good can then be fostered for real good personal development, instead of giving it away to false collective aims currently propagated by media oligarchs.

The educational and media efforts that were undertaken to split away men and women over the last 7 decades were tremendous, so one may be surprised that we still get children at all. Divorce rates are at around 40%, which certainly is unacceptably high. But it also means that 60% of the marriages remain. The glass is more than half full (and less than half empty).

Currently, legal frameworks and the socioeconomic regulations around marriage or permanent relationships make traditional heterosexual partnership between men and women to an existential economic. The determining conditions need to facilitate pairing between men and women making the formation of families with children the societal norm again.

The Insanity Acceleration since 2020 Makes Me Optimistic

As we could see in the population diagrams, the Battle of Sexes was close to finishing off our people.

The escalation in the Battle of Sexes over the last decades was slow, and hardly noticeable. If this slow, hardly noticeable split between men and women and disadvantage of heterosexual love, families and children would have continued at the same non-palpable speed, the end of our people would have been sure.

On a global scale, the war against humanity, beauty, nature, love and life would have been lost to evil forces, too. Since 2020 the totalitarian-suppressive forces became visible and by now normal humans feel stalked and molested by the "bolshe-woke" virtue dictatorship that has infested media and politics.

The Global Power Apparatus is an Illusionary Giant in Decline

The Transhumanist Agenda is pushed by tiny, but unfortunately very powerful groups; plutocrats with their clan structures, foundations and string puppets in the political arena. The fact that such as tiny minority can suppress all of mankind is only possible due to the power of money. Systems of power do appear much stronger than they actually are.

The de-facto power lies in the hand of the many, who are subordinated to the power of the few. Especially in states and systems of power that pretend to grant freedom of speech, the media and opinion manufacturers are of particular importance: If the subordinates may speak out their opinion, those in power need to make sure that the opinions forming among the people are in line with the interests of the powerful. The job of the media is to guide public opinion in a way that induces mechanisms for disciplining dissidents. Dissident opinions will then be punished from the people through individuals, who are in line with the manufactured majority opinion. Such punishments are defamation and social ostracization.

For splitting people into groups, setting them against each other and initiating wars, it is required to control mass media and educational indoctrination systems. Until around 30 years ago, this control could be monopolised in very few hands of rich oligarchs. The advent of the internet makes this tether of control gradually slip away and fray out. Thirty years ago, controlling a few news agencies, newspapers and broadcasting channels was enough to more or less control the narratives and kill stories that could contain truths not appreciated by the powerful. The current tightening on censorship and repressions against free journalists and alternative media would not even have been necessary thirty years ago. The fight of the establishment against "right" (wing) and hate speech appears increasingly desperate. They are losing control. The paste is out of the tube.

Since 2020 more and more of the subordinated people became aware of their subordination. A collective awakening is under way. The massive shattering of trust is perceptible, meaning the end- of governing for the current governing persons in the long run.

I intentionally write 'is perceptible', as the shattering of trust may not have become cognizant; many people may even keep on denying it. The insight that you cannot trust those in power may shake your embedded basic sense of trust.

It certainly is frustrating to see how many people still (want to) trust the suppressive structures in place and for example accept the injection of an inoculant that was shown to do more harm than good. Many of these people still live in a state of cognitive dissonance: They could actually already see that they submit to a dark and evil power, if they would simply be willing to do so.

But remember the Kurt Tucholsky quote "the people do understand most issues wrong, but feel most issues right": The people more and more feel the threat and will gradually rise in their longing for freedom. Soon, humanity, beauty, nature, love and life will find their way and then we will also have children, again.

Wir haben so vieles geschafft - wir schaffen das! (we accomplished so much, we'll make it!).

References

Blanchflower, D.G., and A. Bryson. 2024. 'The female happiness paradox', *Journal of Population Economics*, 37.

Burkhardt, A., W. Lang, and N. Schwarz. 2023. 'Vom Stachel im Fleisch - Wie das Corona „Impf"-Spikeprotein Schaden anrichtet. '.

Campusjäger. 2021. 'Berufseinstieg mit 30: Bloß keine Torschlusspanik! https://www.campusjaeger.de/karriereguide/karriere/berufseinst ieg-mit-30'.

Eicken, J., L. Lindemann, and G. Schulenburg. 2004. 'Schneller Erfolg bei Bevölkerungspyramiden mit GIZEH', *Statistik und Informationsmanagement*, 8.

Fruchtbarkeitskurve. 'https://fertila.de/blogs/kinderwunsch-frau/alter-und-fruchtbarkeit-die-fruchtbarkeitskurve'.

Herman, E.S., and N. Chomsky. Vintage 1994, Erstveröffentlichung 1988. 'Manufacturing Consent: The Political Economy of the Mass Media'.

Kill, N. 2021. 'Faktencheck: „Es gibt kein Volk": Zitat von Robert Habeck wird aus dem Kontext gerissen https://correctiv.org/faktencheck/2021/10/06/es-gibt-kein-volk-zitat-von-robert-habeck-wird-aus-dem-kontext-gerissen/'.

Levine, H., N. Jorgensen, A. Martino-Andrade, J. Mendiola, D. Weksler-Derri, I. Mindlis, R. Pinotti, and S. H. Swan. 2017. 'Temporal trends in sperm count: a systematic review and meta-regression analysis', *Hum Reprod Update*, 23: 646-59.

Markhoff, E. 2020. 'How bad is the bug? Assessing the lethality of SARS-CoV2 in spring 2020 by comparing the age distribution of deceased SARS-CoV2-positives with the age distribution of all deceased from the mortality register in Italy, Germany, France, Spain and England. https://pandata.org/how-bad-is-the-bug/', *pandata.org*.

Mausfeld, R. 2019. 'Angst und Macht. Herrschaftstechniken der Angsterzeugung in kapitalistischen Demokratien. Westend Verlag.': 86-87.

Schmitz-Veltin, S., and A. Schmitz-Veltin. 2011. 'Pfalz. Michael Müller Verlag. 2. Auflage.'.

Statistisches Bundesamt. 2020. 'Geburtenverhalten im Wandel. https://www.destatis.de/DE/Themen/Querschnitt/Demografisch er-Wandel/Aspekte/demografie-geburten.html#:~:text=Im%20Jahr%201970%20war%20dagegen, sogar%20erst%2022%20Jahre%20alt.'.

Steinmeier, F.W. 2020. '„Wir Glückskinder in der Mitte Europas". Rede zum 30. Jahrestag der Deutschen Einheit in Potsdam am 3. Oktober 2020. https://www.bundespraesident.de/SharedDocs/Downloads/DE/P ublikationen/201003-Wir-Glueckskinder-in-der-Mitte-Europas.pdf?__blob=publicationFile'.

Stevenson, B., and J. Wolfers. 2009. 'The Paradox of Declining Female Happiness', *American Economic Journal*, 1: 190-225.

Wikipedia. 2024. 'Liste der Länder nach Anteil an adipösen Personen, https://de.wikipedia.org/wiki/Liste_der_Länder_nach_Anteil_an_ adipösen_Personen#cite_note-who-5.'.